ANOREXIA

The Bare Bones

A first-person account of Anorexia and recovery for frontline professionals

A Tough Cookie Book

Rose Grace Walters

Anorexia: The Bare Bones

Rose Grace Walters

Tough Cookie Publishing

2015

Copyright © 2015 by Rose Grace Walters

The rights of Rose Grace Walters to be identified as the author of this work have been asserted by her in accordance with sections 77 and 78 of the Copyright, Designs and Patents Act, 1988.

All rights reserved. This book or any portion thereof may not be reproduced or used in any manner whatsoever without the express written permission of the publisher except for the use of brief quotations in a book review or scholarly journal.

First Printing: 2015

ISBN: 978-1-326-25955-6

Tough Cookie Publishing

Email: info@toughcookieblog.co.uk

www.toughcookieblog.co.uk

Ordering Information:
Special discounts are available on quantity purchases by corporations, associations, educators, and others. For details, please contact the publisher at the above listed address.

U.S. trade bookstores and wholesalers: Please contact Tough Cookie Publishing at the above listed address.

This book is dedicated as always to my wonderful parents who are with me every step of the way and have never failed to support and encourage me. Also for my fantastic friends and all the people who showed me kindness and compassion during my illness for which I will always be grateful.

About Tough Cookie

I founded Tough Cookie because I recovered from Anorexia myself ten years ago, and after getting through it alone myself, I still see many struggling with little or no support. With so many negative influences online and in the media I wanted to bring something positive and accessible to the table. My experience is specifically with Anorexia and more recently anxiety and body dysmorphic disorder, but I hope to use the things I've learnt (and am still learning!) to help people with all kinds of eating disorders.

I believe in treating people with eating disorders with humility and kindness, and that empowerment and understanding is the only way that they can be overcome. That's why I started Tough Cookie - and why I'm campaigning to end stigma and raise awareness of eating disorders, body dysmorphic disorder and other related mental illnesses from a unique first hand perspective. Through the blog, my books, talks and courses I'm hoping to make a difference and raise recovery rates and address the causes of eating disorders directly.

I'm passionate about promoting happiness, self-worth, kindness and gratitude. I'm also all about **positivity**. I believe that for us to be able to improve the upsetting statistics surrounding Anorexia, it's all about being positive and saying – no, things aren't ideal in mental health services, but you CAN recover and you absolutely deserve to. Angry 'I was left to die' headlines won't get us anywhere – I don't know anyone who has suffered from an eating disorder that hasn't felt let down in some way. We have to look at ways in which we can overcome eating disorders by ourselves as well as accessing the other resources available. I did – and that's why I'm passionate about helping other people who feel let down do the same thing – to make a full recovery and live a healthy, happy life. I understand that people are angry - but anger won't change anything or help others in any way.

I'm living proof that Anorexia can be beaten – even without psychological help. Let's promote recovery in any way possible – with or without conventional therapy, and spread the message that Anorexia is not the end – it's just the beginning.

CONTENTS

1. About this book – *pg13*

2. About me – *pg17*

3. My vision – *pg19*

4. What is Anorexia? – *pg21*

5. Why do people develop Anorexia? – *pg35*

6. My experience – *pg41*

7. My recovery – *pg43*

8. What role can you play in Anorexia recovery? – *pg 49*

9. Treatment guidelines – *pg73*

10. FOCUS – *pg75*

11. The last word – *pg 79*

12. Contacts, recommended resources and further reading – *pg 81*

1. ABOUT THIS BOOK

I wrote this book as an insightful guide for people who work with individuals in all capacities – GPs, psychologists, counsellors, teachers - who may come into contact with people with eating disorders and would like to have a better understanding of the illness and how it feels to experience one personally in order to support sufferers in a positive and helpful way. It's this first-hand knowledge which I hope will help professionals to gain a better understanding and in turn be more able to gain the trust of sufferers and also know how to treat them in the best possible way. All this whilst avoiding some of the damaging things that happen (often inadvertently) which can hinder recovery and do more harm than good.

Why did I write this book?

Following my recovery I wanted to help people to help themselves as I felt I desperately needed a similar book when I was poorly myself, but I also wanted to help others to support them too. Unfortunately I had a bad experience with many of the healthcare professionals I came into contact with, and I wanted to produce a resource which those caring for people with eating disorders could read to help them to understand what is needed and what is and isn't helpful during recovery.

Passionate about positivity and promoting recovery

I know that personally you or healthcare services generally may not have sufficient resources - but that's not what Tough Cookie is about. Tough Cookie is about the small things you can do to help and that starts with a good understanding – which I hope I can provide with this honest personal insight.

For me it's about being positive rather than bashing the NHS and saying recovery is possible, even if you haven't found conventional therapy effective or don't have access to it.

I don't believe that shocking headlines or scary statistics are helpful – instead I'm all about the solution, hope and positivity. I believe that recovery is complex, personal and involves lots of different factors. I also believe that it can be achieved with or without conventional therapies. For me, for us to be able to improve the upsetting statistics surrounding eating disorders, it's all about being positive and saying – no, things aren't always ideal in healthcare services, but you CAN recover and you absolutely deserve to.

The government is not going to find a magic pot of money and if they do, it's unlikely they'll spend it on an overhaul of eating disorder services, so I encourage individuals to focus on other ways to help people to recover and spread the message that even without professional intervention, it's possible to beat Anorexia. Most importantly, I think we can all play a part in recovery, and that starts with a better understanding of how to effectively care for and treat anyone who does come into contact with medical professionals, at any level. Sharing useful information and a personal perspective so that when a person does come into contact with healthcare professionals they are better understood ad better treated as a result because that person can appreciate on some level what they must be going through. This is especially important in the case of eating disorders, because they need a very specific kind of treatment and support to be overcome.

By sharing my experiences in a positive way I'm sure that each and every one of us can make a difference.

2. ABOUT ME

I'm a survivor of Anorexia who is dedicated to supporting people with eating disorders and promoting positive body image. I'm passionate about empathy and understanding, positive reporting and promoting the recovery and prevention of eating disorders through my own experiences, as unfortunately I didn't have access to the help I needed like so many sufferers.

I believe in treating people with eating disorders with humility and kindness, and that empowerment and understanding is the only way that they can be overcome. I'm campaigning to end stigma and raise awareness of eating disorders, body dysmorphia and other related mental illnesses from a unique first hand perspective.

After my eating disorder, I really wanted to make a difference – whilst I wanted to focus on individuals, I was also keen to 'help others to help others' as a result of my poor treatment. I feel that there is still a lot of stigma and misunderstanding surrounding what is a devastating mental illness, and if we can change that where it is needed most, we may be able to raise recovery rates. By sharing my own experience in this way I hope to be a part of that.

3. MY VISION

According to statistics, Anorexia still has the highest mortality rate of all mental illness – a statistic which I first heard about ten years ago when I was going through my own eating disorder. That means that in ten years, we've made little progress to tackle the issues surrounding eating disorders, the things that cause them in the first place, and the way in which we treat them and the individuals who suffer from them.

Whilst most of my efforts are concentrated on general awareness and prevention of Anorexia, I'd like to see eating disorders treated with more urgency, empathy and better understanding. Nobody stigmatised or discriminated against because of their mental illness. Fewer people developing eating disorders – and even fewer people dying from them. It's why I do what I do – and why I want to share what happened to me with you so openly and candidly.

We all have a role to play in prevention – schools, society, family, parents - but when Anorexia does occur we can also do more to increase chances of recovery.

4. WHAT IS ANOREXIA?

As a professional, especially if you're working in the medical or psychological sectors, you might already have a good understanding of what the definition of an eating disorder and in particular Anorexia is. The purpose of this book is not to offer general information or to patronise you – it's to share my perspective and experience. The definition I give below includes my personal view – which I hope might add something to the experience you already have or the opinion you currently hold of what Anorexia is and how it affects an individual – especially someone in their teens.

What is Anorexia?

A person with Anorexia aims to keep their weight as low as possible – even when their weight and BMI are dangerously low and seriously affect their mental and physical health. This involves self-starvation, limiting general food intake, refusing to eat certain foods and sometimes refusing to drink. People with Anorexia will often also exercise excessively to 'burn off' food, even when they are very weak. The effects of Anorexia long-term can be devastating, with brittle bones, infertility, severe hair loss and digestive damage sometimes caused when someone is underweight for a prolonged length of time.

Anorexia and Bulimia can be present together or can feature elements of one another – for example, a person with Bulimia may also exercise excessively, and a person with Anorexia may sometimes purge to get rid of food they have been forced to eat. Predominantly eating disorders affect women, but men can and do also develop eating disorders.

Both Anorexia and Bulimia are very serious conditions and often arise out of a need for control. They can be caused by the sufferer feeling 'fat' and wishing to be slim or to lose weight, but that's not always the case. Anorexia used to be known as 'the slimmer's disease' – and it's this and largely false media that has led people to believe the misconception that eating disorders are all about food and body image, when in fact there are many other causes and triggers behind them. There is more on this in **Chapter 5**.

Symptoms of Anorexia

Symptoms of Anorexia include:

- Eating very little, avoiding fatty, sugary or high-carb foods, obsessively controlling food intake
- Missing meals and avoiding meal times
- Not allowing others to prepare food – or needing to know what is contained within meals cooked by others
- Anxiety and stress over eating out or situations in which they have little control over what they eat
- Lying about having eaten or hiding food

- Obsessively counting calories in food and an obsessive attitude around food and eating, including the scrutiny of others' eating habits
- Irritability – especially when questioned about food or their behaviour
- Repeatedly weighing or checking in the mirror – with unrealistic expectation and perspective of their body size and image
- Looking physically thinner
- Weakness, dizziness, hair loss and dry skin - symptoms associated with the physical side effects of starvation
- Isolation – losing interest in doing the things they used to love or seeing other people

Anorexia can be accompanied by other mental illnesses, such as OCD, anxiety, depression, self-harm and low self-esteem as well as alcohol and drug abuse.

My experience with Anorexia

There are so many misconceptions surrounding Anorexia that I wanted to dispel a few myths and give an insight into what my definition would be, as someone who has experienced it and recovered. I go into some detail in this book because for someone who has never had Anorexia but is charged with treating or looking after people who do, I think it's important and necessary to know how it works, how it gets into your head and what happens in the early stages so that you can be more aware and look out for any warning signs which might be missed without inside knowledge.

A friend of mine who is currently suffering from Anorexia echoed my sentiment when she referred to magazine articles featuring celebrities and 'real life' sufferers who were now 'beautiful and successful.' The message she got – as I did, before I developed Anorexia, was that Anorexia was a solution not a problem. A method for obtaining everything I wished so much that I had – beauty, popularity, financial success. Those things consumed my mind - I've now realised that there are other things that are much more important in life, but at the time I was really only a child, a teenager at most - and because I was being bullied I desperately wanted to be liked. Those things appeared to be a sort of 'magic formula' for popularity.

If I could lose a lot of weight, all I would need to do is put a little back on, so I wouldn't be the person I was before (who I believed was wrong in every way possible, disliked by everybody and 'fat'), but I would instead be healthy and slim and beautiful and smiling in a magazine talking about my journey and being praised by others. This is exactly how it looked to me – but of course it wasn't like that at all. I had an ignorance and lack of understanding of the illness, as so many people do. I didn't realise that it wasn't something I could control – and in fact it took me years to realise that it was in fact controlling me, and not the other way around.

The signs of an eating disorder are often quite subtle, but can develop rapidly and quickly increase in intensity over a short period of time. Watch out for the signs such as secretive behaviour, a lack of desire to eat around others, not wanting to go out to dinner or be in a situation where they cannot control their food, and of course the obvious symptoms such as getting thinner and excessive exercise. Eventually these symptoms are so intense that it is impossible to misconstrue them for anything other than an eating disorder.

You become so obsessed with food that it consumes you; it's all you think about. You have no time for anything, or anyone, else. You become a shell of a person, you become angry and irritable. As you eat less, your brain (as well as other major organs) stops functioning properly. That means it becomes harder to think clearly, resulting in you becoming more and more irrational. Logic, rationality and ability to reason fly out of the window.

I have always described Anorexia (I can't speak for Bulimia, however I imagine it is very similar) as a demon inside my head. I was still there; my personality, my humility, everything I loved in life. Yet slowly I had been replaced; my brain had been taken over by another entity whose sole focus was to destroy me. To kill me, ultimately. This demon told me I was better off dead. It convinced me that I was disliked because I was fat and ugly and anything other than my current state (morbidly underweight) constituted that (fat and ugly); therefore I could either be unlovable or die. It had to be one of the two – there appeared to be nothing in-between. I was miserable and incredibly poorly physically, but I was controlled fearfully by my illness.

Relatives and friends told me to 'just eat'. And whilst I wanted to, this parasite inside my head strongly advised me not to and I had no way of escaping the fear it had instilled in me of leading a normal life.

Food became an enemy, and along with it so did everybody around me who tried to help. But because the demon got angry with my family and friends, they got angry back. They felt helpless and frustrated and a lot of professional people who were supposed to be helping me directed cruel, venomous words at me or pointed the finger at my family, who were desperately trying to get me back.

My personality had completely disappeared by the time I was at my lowest weight – I was just drifting around, a shell of who I used to be, unable to talk or smile or think properly. The only thing occupying my mind was how I was going to avoid the next time someone tried to get me to eat or drink something. I was fearfully wondering whether the air I breathed contained calories. Very much wishing I could just die - knowing I was nearly there. Inside, the 'real me' was frightened and guilty.

It took a long while for me to realise that I was worth a shot, and to decide I wanted to live a proper life. It was a journey of ups and downs – some days I felt I might be worth recovering for, other days I simply wanted to die. I believe (as I say later on in the book in Chapter 5 about family support) it was constant encouragement from people around me, the little things that they did and said which slowly but surely made a huge difference in my mind. I started to want things that I just couldn't have if I weighed just a few stone and lived in a hospital. It was the most difficult thing I ever did but I look back now with immense gratitude for having been able to come through it, not without lingering scars and difficulties, but without relapse and in a position to be able to help others.

How I describe Anorexia

I describe Anorexia specifically (as this is what I have personal experience of) as a parasitic demon because for me it was almost like being trapped inside my own brain with something which wanted me to die and convinced me that I was better off if I followed its instructions without exception.

I realised when I was doing final edits on my other book, Tough Cookie, that every time I talked about Anorexia I had personalised and personified it, and kept referring to it as another human being. I think this is because that's almost how it was for me – like having another person in control of me inside my own head – a person who I couldn't ignore because they seemed to make a lot of sense.

Anorexia can be schizophrenic in that sense – a loud, vicious voice in your head which has a personality all of its own – a personality which terrifyingly comes out in the words you speak and the things you do. My mum likened it to me being possessed when I said certain things which were completely out of character or screamed at her calling her a bitch and telling her I hated her because she was trying to help me by making me food. I don't think many people see this side of Anorexia – or talk about it – which is why I think it's important to share it in this book.

Common Misconceptions

Whilst we are all more aware of eating disorders, many who haven't experienced one themselves or have had the misfortune to watch someone they love battle one have a warped perception of what they are and therefore have no idea what to think about them or how to deal with them – even if they work in the medical profession. A friend of mine with Anorexia told me recently that whilst she was working at a hospice, many members of staff she encountered told her that they simply couldn't contemplate working with people with eating disorders or other mental illness because those people were unimaginably 'selfish' compared to the patients they currently treated who were dying 'through no fault of their own'. This over-emphasis on the personal responsibility of people with mental illness, but especially eating disorders and Anorexia, is something which I believe really harms the chances of more people becoming well again. Stigma and judgemental assumption held by people, especially those who are supposed to help, is damaging enough to cause many never to recover. I was subject to this way of thinking whilst I was poorly – from the staff who were charged with looking after me, relatives and total strangers. For a 14 year old with Anorexia that was devastating - and the way I was treated has undeniably stayed with me. I share some common misconceptions here to help you to understand how it feels to have an eating disorder but also to demonstrate some of the misguided beliefs lots of people hold about what an eating disorder is and how someone with Anorexia should be treated.

'An eating disorder is a choice'

A common misconception about eating disorders is that the person involved has a choice and has wilfully made a decision not to eat. It's simple – it's solely about food and the person's apparent choice not to eat it. When I was poorly it didn't occur to many people that Anorexia wasn't a choice – just as cancer and MS aren't choices. It is a very difficult thing for people to get their heads around – and I was often attacked by family members, medical staff, even strangers - which is why I do what I do with a goal of using my own experience to explain Anorexia to others.

Eating disorders are not self-inflicted; it is not a choice to have an eating disorder. Because it masquerades as the individual in question, it's easy to come to that conclusion. The words come out of their mouth; the actions are made by them. But what I and others who have had or currently have an eating disorder know is that it's not really you, and it's very hard for others (and sometimes for you yourself) to differentiate between what's you and what's the eating.

Nobody should **ever** make anyone feel guilty for having had an eating disorder, just as it would be unthinkable to accuse somebody with cancer or MS or Parkinson's of 'ruining someone's life' or being 'selfish'.

'Eating disorders are self-indulgent'

When we believe that sufferers have a choice, that's when we conclude that they are clearly doing what they are doing for some selfish reason. How *selfish* that person must be to put their loved ones through so much pain – to take up valuable healthcare resources, when they could just eat.

Far from a cry for attention, eating disorders are dangerous, complex and much more serious than a refusal to eat.

'Eating disorders can't be overcome'

This is **untrue** and I am living proof of that – along with lots of others who have recovered and continue to recover fully without relapse. However the numbers of people achieving this are way too low.

I believe that when we focus on conventional methods and discount everything else, we close the door to recovery for so many people. This combined with a lack of understanding is what makes Anorexia so deadly. But if we spread understanding and encourage recovery through any method possible, we can make a difference.

In Tough Cookie, I talk about conventional therapy, but I also stress that if this 'hasn't worked' (for whatever reason – maybe the therapist or timing wasn't quite right, or the therapy wasn't appropriate) – there are alternatives and they *can* still get better because that was my experience.

I include alternative tools and ways of thinking for people struggling with eating disorders in Tough Cookie for this reason – some of which I'll go into more detail about (along with how you can help) in **Chapter 8**.

'Eating disorders are a product of weakness'

Weak people don't get eating disorders. You have to be very strong in fact, I believe, to carry something like this on your shoulders. Invariably sufferers go on and on and on – even those who sadly eventually pass – for months, years in an emaciated state. Some go to work, go about their daily lives, hiding what's going on for unimaginable periods of time. These are not the actions of weak people. These are the actions of people who are poorly beyond belief, consumed by an eating disorder, who desperately want to be right in some way but feel all wrong.

'Eating disorders are all about food'

People tend to think that an eating disorder is all about food – the person just 'won't' or 'can't' eat and they need to be encouraged to eat again. This is probably because it heavily involves food and also involves the person losing weight. But an eating disorder is not just about food – it's so much more than that. It's not about being faddy, awkward, stubborn or being on an 'extreme diet'.

Food is the product – unusual habits around food are the resulting behaviour, so it's easy to assume that food is the root. But actually the root can be a number of things. Low self-esteem, negative beliefs, OCD, anxiety and depression can be behind an eating disorder. To give an example, the root of my eating disorder was pretty much all of the above plus natural perfectionism. Because I was bullied for being fat and ugly day in day out at such a young age, I believed that – so I also believed I was a bad person, not worthy of other people's affection or a life like anyone else's. When I went on a diet I did so to lose weight to be liked, but I soon became disillusioned with my slow progress and when I saw eating disorders glamourised in the press I thought that restricting my food intake must be the way forward – because then I could be like these popular celebrities. I soon became 'addicted' to losing weight and a few favourable comments from the bullies at school spurred me along. So although the result was Anorexia, the root cause was that I was very unhappy inside – I hated myself.

Once you realise that an eating disorder is not a choice, everything I've discussed above should seem obvious. Please never make the assumption that someone is weak, selfish, inconsiderate or stupid because they have an eating disorder – and if you hear people saying otherwise, correct them using the things you will learn reading this book.

5. WHY DO PEOPLE DEVELOP ANOREXIA?

Of course because we are all individuals, we all have different motivations and catalysts behind us developing an eating disorder, contrary to some common beliefs. As I say previously however, it's not at all about food, or how you look – it's about how you feel inside.

I think more than anything pressure is often involved. Pressure to be perfect – whether that's aesthetically, academically, personality-wise or in our career. As a society we are much more aware of what other people are doing via social media and the internet and that causes us to compare, often unfavourably.

For me, it was a combination of those things, very low self-esteem and family trauma which led me to develop Anorexia. No matter what you might think initially, always consider that there may be (and probably are) multiple causes behind someone's condition and that it's likely they have already been through some sort of considerable difficulty and emotional distress before they reach this point.

Everyone is different

The first thing to say when talking about causes and triggers behind an eating disorder is that everyone is different – so naturally from one person to the next the reason behind the illness will be different. It's difficult to be able to place the 'blame' for an eating disorder at the feet of one thing in particular, because it's often a combination of things which results in the illness rather than one issue alone. There are triggers –for example a traumatic life event such as the death of a loved one, or internal issues which build up over time as a consequence of bullying or sexual abuse. It might be that there's nothing blindingly obvious - just a feeling of discontentment which has been bubbling away for months or even years and slowly leads to the eating disorder. Eating disorders are also often accompanied by other mental illnesses - for example, I had OCD and Body Dysmorphic Disorder too. Others might have Bipolar Disorder, Depression or Anxiety. Recognising this can help people to understand how complex an illness this really is - especially when it does come along with other complications.

Media and culture DO have roles to play

I once read an article written by a top psychologist claiming that society's obsession with 'perfect' and in particular 'the perfect (slim) body' had absolutely nothing to do with the development of eating disorders, in particular Anorexia. As someone who went through Anorexia, I have to disagree.

I developed Anorexia for many different reasons - but the root for me was being bullied for being fat. I decided (with the help of the media and kids at school) that thin people were liked, and embarked upon a journey of fad dieting which ultimately led me to become completely obsessed and very poorly. Whilst my self-hatred, perfectionism and lack of control over my life certainly had a lot to do with my illness, with the correct nutritional advice, kindness and encouragement from responsible sources I might have better understood how my body worked, respected it rather than hated it, and been able to see the world with a more rational perspective. I may have developed a healthier level of self-worth, maintained more positive mental wellbeing, lost the weight I needed to lose safely and with a new-found self-confidence I might not have cared so much about what people at school said about me. Instead I was influenced by celebrity magazines and dubious information from diet companies. I was taken in by adverts and TV programmes which showed me who I 'should be' and clearly highlighted who I was not and why I didn't fit in.

Pressure in today's society isn't purely aesthetic, either. Pressure to excel in every area of our lives is becoming a huge trigger for eating disorders, as is a life increasingly lived on and through a screen. We're exposed to other people's lives more than we ever have been before, and this in turn encourages us to compare, often unfavourably.

Perfectionism and control matter

I am naturally a perfectionist and I like to be in control. This is in part a personality trait, but also perhaps a product of what I have been through in life. Several studies have shown that people who have these personality traits can be predisposed to developing an eating disorder. People with OCD and generally obsessive or anxious personalities (like myself) can also be susceptible to eating disorders by nature.

Food ends up being the resulting behaviour - because food is easy for us to control. The root is self-loathing, a feeling of total lack of control, a desire to be 'perfect' and a strong determination to reach goals. And when goals are not reached, a punishing, critical voice inside which scrutinises everything which could have been improved upon and berates you internally. All of this in turn can of course affect how you behave externally. As above, it's different for everyone.

Statistics show that more and more people are developing eating disorders, at increasingly young ages. I think largely this is down to social pressure – now as adults and as children we are expected to be clever, good looking, funny, popular, interesting all at once – the list goes on. Society focuses on what we haven't got and places large amounts of value on material things without considering what's on the inside. Schools and colleges place huge academic pressure on children at early ages, with targets and grades and league tables taking precedence over the mental health of our young people. Bullying is more prevalent than ever before with an increased number of avenues for bullies to exploit, whilst social media encourages a culture of having more or 'needing' to have more.

Perfectionism has always been a part of me and certainly contributed to my development of Anorexia – but now, anything less than perfect is rarely acceptable. 'Perfect' is depicted as attainable by celebrities, social networks and other media, so naturally younger people feel pressure to be perfect, look perfect and have a 'perfect' life. Of course perfect doesn't exist, so many children and young people feel disappointed and ashamed when they don't live up to society's ideals. When surveyed, UK children were amongst the unhappiest in the world, coming 14th out of 15 countries including Ethiopia, Colombia and Romania for overall life satisfaction – and as external pressures, perfectionism and bullying were causes behind my own eating disorder I believe this statistic must be intrinsically linked to the rise in eating disorders amongst children, teens and younger people.

I think more than anything **pressure** is one of the elements most commonly involved in the development of eating disorders in young people. Pressure to be perfect – whether that's aesthetically, academically, personality-wise or in our career. As a society we are much more aware of what other people are doing via social media and the internet and that causes us to compare – often unfavourably.

For me, it was a combination of those things and family trauma which led me to develop Anorexia. No matter what you might think, always consider that there may be (and probably are) multiple causes behind someone's condition and that it's likely they have already been through some sort of considerable difficulty and emotional distress to reach this point.

6. MY EXPERIENCE

Unfortunately, I didn't have a very positive experience with healthcare professionals whilst I had Anorexia.

Although my GP was very helpful, initially she struggled to find resources I could access because of my age, and I wasn't eligible for a referral until I was in dire straits, so I had to deteriorate significantly before I was finally seen by the CAMHS team at my local hospital. I was seen by a number of therapists but one in particular was assigned to me – a man who didn't understand me and insisted on telling me how I felt without listening to me properly, which alienated me from him and eventually made me hate him so much that I refused to see him. As his input was far from helpful my condition worsened, and I was admitted with some acute medical difficulties. At that point I was eating as little as possible but I was also hardly drinking, as I believed water had calories in it.

The time I spent in hospital, rather than healing me mentally and physically, caused me to deteriorate in both ways. As I was admitted to a general children's ward not a specialist unit I was given the only resources they had to hand – a dietician with no knowledge of eating disorders, psychologists with no understanding of eating disorders, and doctors and nursing staff who had never seen anyone like me before. The result was a difficult experience which was incredibly distressing for me.

Lots of bad things happened to me whilst I was in hospital, and I wasn't treated well at all by staff at a number of levels. But I don't believe that it's helpful to go over everything here in the book, as whilst I experienced very bad things mostly, I also experienced a few very good things. Looking back on these good things has helped me to see how I found my way out of Anorexia - even though I didn't get the treatment I needed. I talk about these in **Chapter 8** when I discuss what is helpful – so although my experience was mostly negative, it wasn't *all* negative – and I still recovered despite the negligence I suffered which is actually very positive. Of course, I don't want anyone else to go through what I did – which is partly why I wrote this book.

7. MY RECOVERY

Recovery can be a frightening prospect when you're suffering with an eating disorder.

When I was at my lowest, I really wanted someone who had had Anorexia and had 'come out the other side' (as it were) to come and sit with me and tell me everything was going to be okay. That it was worth getting better – that *I* was worth getting better. What had it been like for them, what might it be like for me? All these questions went unanswered.

Surrounded by people who were supposed to be looking after me but didn't seem to care or understand, I wanted someone to do just that and support me. Luckily my parents had access to a therapist who helped them to help me – which again shows the power of your contribution in recovery just by supporting family members and those around the individual in question.

What I aim to do by talking about my recovery in this book is to give a clearer idea of what it's like to go through recovery and the triggers which can cause someone to worsen or to improve psychologically and therefore physically, too.

My Recovery

Over the years, lots of people have asked me: 'How did you recover from something like that?' It seems that whilst understanding of eating disorders is slow to spread, it's still recognised that they are one of the most dominating, most dangerous and most difficult mental illnesses to overcome. It's taken me a long time and a lot of thinking to come to a conclusion as to what the catalyst was for me making what appeared to be a snap decision to 'get better'.

Family, friends and medical professionals had all spent a lot of time urging me to listen to them and 'stop' what I was doing. Having an eating disorder was often presented to me as a choice by healthcare professionals and loved ones alike – it was my own obstinacy and wilful determination which was driving me to do what I was doing. I might have appeared to be in control of the situation, but in fact I was helplessly being dragged along by my eating disorder. My parents often pleaded with me to understand that I needed to be well again, that I should be well again, and that they desperately didn't want to lose me.

Whilst when I really thought about it, I could recognise that my life was far from ideal, at first for me recovery represented returning to an equally awful life which I had come to despise the more I thought about it and the more fixated I became on becoming invisible and disappearing altogether from the world. I failed to recognise that in fact, there had been a lot of good things in my life before I'd become poorly, things I'd overlooked because I had become so engrossed in everything I felt was wrong with me.

I left hospital weighing only a little more than I had been admitted, and with no psychiatric improvement or resources to help me to cope at home. Because I'd been on a children's ward and looked after by staff who didn't understand my condition, I'd been allowed to go for hours without observation. I'd been going to the toilet, having showers and brushing my teeth alone with the door locked.

It took weeks and weeks of fighting and screaming and crying over mealtimes and long, intense days which nearly drove my parents to breaking point before I started to even think about recovery. But eventually I started to feel guilty. I started to think about the things I loved and enjoyed, and I started to contemplate the future.

Having a bath is horrendously painful when you are only made of bones. With no muscle or fat to cushion their fragile, protruding joints, my bones rubbed painfully against the hard plastic bath tub. I was bruised and brittle, my skin yellow tinged with red and black and blue at the extremities. I was so cold, cold to the core, even in the boiling hot water. I have a mirror on the wall in my bedroom and I'd not paid a lot of attention to the person standing in it fully for a long while; I was so focused on the elements of myself I found least attractive and tracking their 'progress'. However in the dim light coming in from the hallway I now saw the most horrific, skeletal figure standing before me. I was unrecognisable. My thinning short hair was hanging wet and limp over what could only be described as a scarily-fragile looking skeleton covered in pale skin. My face was gaunt and my eyes sunken, lips cracked, my chin pointy. I was just bone. I remember thinking only: 'What have I done?'

After that I started making a conscious effort to fight the voice in my head. I started eating more and eventually reversed the determination I'd been using to starve myself, using it instead to put weight back on. It wasn't that simple – of course after that initial moment of clarity I struggled with relapses and my journey to becoming healthy again was long and slow. My decision to recover seemed epiphany-style, but it wasn't - it couldn't have been, there must have been a build up to that event, little things that influenced me over time to decide 'all of a sudden' that I wanted to recover.

Partly, I had put a little tiny bit of weight on, so I could see with more clarity what I had done. When the brain is starved, logic and rationality go out of the window. My body dysmorphia had masked how I really looked previously to the point where I was so obsessed with my legs, stomach and cheekbones that I saw nothing else and sometimes I was pleased with how they looked, but at other times I beat myself up or panicked because they weren't as I thought they should be. I'd blocked out all the pain – just like I'd blocked out the hunger previously. Just like I was effectively blind to how things really were, and how I really looked. I wasn't living in reality, but instead I was trapped in my own version of it. Or rather Anorexia's version of it. Yet now I could see it all and whilst Anorexia had far from loosened its grip completely, the 'me' inside was wanting to recover and starting to fight back.

Mostly I attribute my recovery to the love and kindness I was shown by my parents. They worked hard to show me that life wasn't just about food. They worked hard to show me that I had a life outside of what was going on right now – to remind me of my likes and dislikes, my hobbies and interests. They spent time with me and treated me like a human being (especially when it mattered the most) but equally when the time came they were unfalteringly strong and by being uncompromising and defiant in the face of what is every parent's worst nightmare they told the Anorexia where to go. At the time, it felt to them like they were punishing me. It seemed like I was the one calling them names and being unkind. But deep down they knew it wasn't, and they made a big effort not to take the massive strain of my illness out on me.

Through this care and love, they helped me to realise that I wanted to live, and remember that there were lots of good things about life that I loved and enjoyed.

It certainly wasn't a case of making up my mind to be better and then setting about it and that was the end. Recovery was a long, slow process which had peaks and troughs. I found myself adopting old habits and relapsing without even realising until it was brought to my attention. I often found myself controlling my food and that of everybody else to an unreasonable degree. It took a lot of time and it was physically a hard slog to put the weight back on that I had lost and try to repair the damage done to my body.

I look back now and know I'm incredibly lucky to be alive. I've had experiences and hardships since which have almost been on a par with my eating disorder (I'd say it's probably the worst thing I've ever gone through, but because I was numb for much of it I've certainly *felt* worse since), and it's been really difficult for me ever since because my issues went unresolved. But despite this I can honestly say that choosing to recover is the best thing I have ever done. I've had a fabulous life so far, and I have done so many things. Most of all, I have been able to help others.

That demon does not belong there – it is taking up the space which is meant for that person's life, their dreams, aspirations, hobbies, relationships. It will tell them that none of those things matter but when they are encouraged and shown they can find the will to live again inside of themselves.

For me, recovery is all about remembering who you are and what you love, which in turn helps you to want to live again. I think it's important to combine a strict psychological and medical approach with humility, kindness and care for the person who is still there, almost trapped inside their own body with this unseen entity.

Although I share this to give an idea of what recovery is like from a personal perspective, it's also important to note that my recovery didn't happen in the way it should have done. I should have had proper psychological support and nutritional advice so that I didn't stuff myself to gain weight, and I should have been monitored to ensure that I didn't have any adverse effects as a consequence of eating so much food so quickly when I was so underweight and had been so poorly. Lots of people also go through recovery alone as I did with my parents, so it's really important to recognise this and if you do come into contact with anyone who you feel might need extra support, please try to advise them in the best way possible or find other resources they may find useful.

8. WHAT ROLE CAN YOU PLAY IN ANOREXIA RECOVERY?

Your role in recovery when you come into contact with someone with Anorexia will largely depend on your position and the capacity in which you treat the individual. With this in mind the guidelines below are general – some may apply to GPs, some to nurses, some to HCAs and some to doctors, or just one of those parties individually.

Whilst we do need to identify where things go wrong, I also think it's really important to recognise what went *right* in my case in particular. Encouraging and praising individuals for small steps (however small) and recognising the people who ask for help is going to be more powerful than simply talking about the things which were clearly not helpful or fair in lots of detail. It also demonstrates that little things can make a powerful impact and a big difference.

There were certain things that happened to me throughout the course of my time in hospital that I now realise constitute abuse – I share these here not to penalise the NHS and those involved but to give good examples of how not to treat sufferers, and how the thoughtless (often unintentional) actions of others can prove to be very harmful and cause lasting mental damage.

What isn't helpful?

Lack of urgency

It's a **huge** first step to ask for help – so when someone agrees to get help and is then turned away it's devastating.

There are lots of reasons why people with eating disorders are sent away, or why their conditions aren't recognised. Sometimes it's simply because the GP has no knowledge of what to look out for or doesn't know what resources might be available for that individual – at others it's a wilful denial because of misconceptions like the ones I speak about in Chapter 3. Sometimes, it's simply down to a lack of resources – no beds, no places for therapy – then the hands of GPs, teachers and other professionals are tied.

The biggest first step is recognising and admitting you have a problem, and then asking for help – so to deny someone that help is pretty soul-destroying. As an adult, you have to make the choice to be in the GP surgery, at the hospital, or in treatment of any kind – so it's important to recognise this as a massive achievement wherever they may be.

Someone who couldn't eat because of a purely physical problem and who was on the brink of death, organs shutting down, their body wasting away, would not be turned away from hospital. But if it's a mental health problem and you have an eating disorder despite the physical danger you are in you are sometimes neglected. This has to stop. As discussed, there is still a way of thinking about eating disorders which involves an element of 'choice' – just as alcoholics 'choose to drink'. It's not a choice. These people need help.

So whilst I recognise and understand that resources are scarce – try to do the best you can do with what you have and always praise the person for small things like trying to access help or admitting they need support - your recognition of their struggle may just spur them on to work a little bit harder to get to recovery.

When I was consistently turned away from services because of my age, and then treated in the wrong type of hospital with the wrong type of staff and psychiatrists with no interest or knowledge of my illness, it made me feel like nobody cared and more than that, I was right to be doing what I was doing and could carry on. I already felt like I wasn't worth fighting for – I didn't need their help to feel bad about myself. The people I met and the things they did have stayed with me – good and bad. It doesn't have to be a grand gesture – just a kind word or taking the time to listen and understand how that person might be feeling. You'll hear about some of the small acts of kindness I received from strangers and staff whilst I was poorly later in the book, and they have stayed with me and really brought me up a little when I was at my lowest.

Each week and each day that goes by is critical with an eating disorder. Day by day you lose more weight, become weaker, more poorly. Delays and denial of access to services *feeds* the eating disorder.

Noticing signs of this secretive illness and how to treat it – whether you're a GP, youth worker or charity volunteer – is crucial.

Blame

My mum had just lost her dad (and I had lost my Grandad) not long before I started to become poorly. The person who later became my therapist at CAMHS called her and proceeded to berate her for not having accessed the service sooner – apparently unaware that my mum had been trying to get me the help I needed for *months*. My mum still feels guilty as a result and I could never forgive him for hurting one of the people I love the most – a person who gave me a lot more psychological and emotional support than they did!

An eating disorder isn't anybody's 'fault'. If I was looking for someone to blame then I'd have to point the finger at my bullies, but that doesn't solve anything. Whilst identifying a cause is important, going over past events and trying to find an individual to blame is fruitless and can actually be quite harmful for the individual and the family involved.

Not being listened to

Because I was 14, I was treated with a complete lack of respect. I was patronised again and again until I stopped feeling like I could be open about anything. I never got to the stage in therapy where

I discussed being bullied, needing control, my OCD, my perfectionism; therefore the root problems were never addressed and I still have mental health problems today as a result.

They were intent on telling me how I felt. There's nothing more frustrating and upsetting than having people making a judgement about you based on their opinion and assuming things about you that aren't true, telling you who you are rather than listening and interpreting.

A complete lack of care

Nobody at a higher professional level who dealt with me in hospital (therapists, psychologists, consultants) ever showed me any sort of compassion. Their attitude was similar to the other medical staff who I came into contact with – and whilst I can understand ignorance on their part I'm still shocked that people trained to look after children with mental health problems could be so uncaring.

A little kindness goes such a long way. A genuine interest in the person's recovery and wellbeing goes even further.

Ignorance and misunderstanding

I've already discussed how misconceptions can be harmful – which is why I de-bunked them to hopefully prevent them being spread both in society and in a smaller medical community. But ignorance can take a more innocent form – through a simple lack of information people can sometimes make the mistake of underestimating Anorexia, perhaps justifying the person's denial of services or delay in treatment because they just don't realise how urgently they need to be seen.

I've discussed in some detail how Anorexia works in secretive, cunning ways – and some of that information may have surprised you. The reason I included it was to share for anyone who hasn't come into contact with Anorexia just how devious it can be – however lovely and trustworthy the person suffering may appear to be. They're not the ones in control – even though it appears that way.

Anorexia is devious and clever. You cannot ever underestimate the lengths it will go to stop food and water going into your mouth. It will hurt your family and friends and see you dead before it stops getting its own way - and that is the unfortunate seriousness of this mental illness.

Because I wasn't watched enough by innocent staff, I continued to do a lot of exercise during my stay in hospital and I was able to squirrel all that food away. I don't think that this was actually neglectful – I think they trusted me and didn't have any understanding of Anorexia which would have made them suspicious or more careful. This ultimately led to me going home in a worse state than I arrived in, with a much better idea of how I could hide food, conceal exercise and avoid eating.

Negligence

My friend is currently undergoing treatment at a specialist unit for eating disorders. She is an adult, so she is accessing adult services for eating disorders. When she arrived in a similar state as I was when I was admitted to hospital, she was immediately put on constant observation, a strict refeeding plan and was monitored every hour to check her overall physical health because of the high risk of death associated with her condition at that point.

I had my blood pressure checked once every few hours – but that was it. I should have had my bloods taken and checked and full observations done every hour, with fluid monitoring and other important checks.

I saw a general dietician who just told me I needed to eat a lot more and sent up a calorie-controlled plan to the ward accordingly. As part of this, I was expected to eat huge portions of carb-heavy foods like jacket potatoes, shepherd's pie, stew and chicken nuggets and chips, when I had only been used to very small amounts of fruit and water for weeks. I've learned since that this could have killed me – as at such a low weight and with certain organs failing I should have been on a proper re-feeding programme which would have ensured my body would have recovered properly at a slow, steady rate . It's well known that giving someone who is dangerously underweight and hasn't eaten properly or taken enough fluid in weeks even a normal amount of food can cause heart attacks and seizures which can easily kill them.

I suffered massive discomfort because my bowel was suddenly forced to process larger amounts of stodgy food – and when I did feel as though I wanted to recover, I did so by eating as much junk food as I possibly could to put on weight. So without the proper advice nutritionally speaking and the psychological help I needed my relationship with food only actually improved and normalised a few years ago – and even then I still struggle sometimes.

No monitoring, no proper refeeding, inexperienced psychologists and dieticians - so many things they did could have killed me, as well as everything I was doing. For this reason I think it's important to ensure that people admitted with Anorexia are properly catered for and treated even when they don't appear to be severely affected – but especially if like me they are at a very low weight and are at risk of complications.

Punishment and abuse

Some of the nurses who looked after me were wonderful. They didn't understand my condition at all – but they didn't treat me as any less of a person because of that. Instead, they showed me concern and made time to talk to me and take care of me as best they could – even though Anorexia was able to get away with murder under their noses.

Unfortunately though there were only a couple of nurses who treated me in this way. I think the staff felt I was in the wrong place (they were right) and I was a difficult, inconvenient patient –

although I was placed in a bed opposite the nurses station, they had to try and watch me constantly and of course had a battle on their hands when it came to meal times. They were obviously under instruction to make sure that I ate what was put in front of me, but some of the less tolerant members of staff took to sitting right opposite me whilst I sat on the foot of my bed weeping over another plate of food, shouting at me angrily and berating me because I wasn't finishing it 'quickly enough'. They were frustrated because they couldn't understand why it was so hard for me to eat: 'It's not difficult, just put the food in your mouth.'

Another nurse felt that I shouldn't be able to eat in the relative privacy of the ward anymore – as this was generally not allowed and patients went up to the canteen, usually with their parents. This meant a humiliating and incredibly upsetting trip to the canteen for me one lunchtime, surrounded by children and babies with their horrified looking parents as I was cajoled, crying, into eating half a banana.

This sort of treatment has stayed with me because I felt sub-human as a result of it. These people were angry with me – to them, it looked like I was being stubborn – they couldn't separate me from my mental illness. They were angry with me because they thought that it was my choice not to eat and didn't see the way I was torn inside wanting to do the right thing (for my parents) but having a voice which viciously told me not to and made me afraid that if I did I'd be hated even more. Anorexia was undisturbed by people's negativity and anger, but I was hurt by it – which understandably was a dangerous and harmful situation for me to be in and made it harder for me to recover. I was in completely the wrong place with the wrong treatment, and as a result I suffered.

What struck me is that even though none of the staff were trained in my mental illness, they all reacted so differently to me. Some were extreme in the way that they seemed to resent me for being the way I was; others were shocked and appalled and just wanted so much to help me because they saw a vulnerable young girl suffering.

Poorly trained staff is a risk for a number of reasons. On one hand, the staff who were abusive damaged me mentally, but the lovely staff also hindered my recovery by simply being too kind and allowing the Anorexia to flourish as well as showing compassion to me.

Telling someone with Anorexia to 'just eat' – that it is as simple as putting food in your mouth, assuming you have control, saying 'you don't have a choice so you might as well' , 'you're not trying' is just like telling somebody with a broken leg to 'fix it themselves'. Mental illness, like physical illness, requires professional attention and that person needs help from a trained expert to 'fix what's wrong'. Anorexia is unique in that it requires both medical and psychological attention – and this is why I think it's often not treated properly as the combination of the two needs to be in place and administered by people who understand the illness in some depth. Ten years after my experience, I believe that there is still a lack of understanding amongst medical professionals and even psychiatrists.

I noticed that in a lot of circumstances, those professionals who are supposed to help you can (sometimes inadvertently) treat you like an illness and not a person. It's important that you feel that staff genuinely care about you rather than feeling they're just doing their job.

From the subtle (limiting my visiting hours, dangling the carrot of being able to go home then retracting it, making decisions without explaining them to me and saying things about me that weren't true) to the not so subtle (shouting in my face, making me eat publicly in front of other children who were frightened of me because of how I looked, constantly questioning why I wouldn't 'just eat') – the bad things that happened to me have stayed with me.

Even if you don't understand, reserve judgement and watch your words. Don't tell them to 'just eat' don't tell them they're not trying – they're not in control.

A lack of resources and knowledge of resources available

My GP wanted to help me, but she really had to research ways in which she could find support for me and it took her weeks to get through to somebody who agreed that she could refer me to CAMHS. I think the reason for this was a combination of the above – and as I've said before, delays can cost lives.

There were adult specialist wards and hospitals, adult referral centres support centres, charities and counselling – as a 14 year old all of those doors were closed to me. Younger people and children are still not provided for as they should be, just as adults aren't, yet we tend to see more cases in people of an increasingly young age.

In an ideal world there would be sufficient good quality outpatient care, therapy and resources whilst people are waiting for beds in specialist units for the treatment they desperately need.

However there's not much we can do about this unless we are politicians, and even then the public spending pot is stretched beyond belief. As individuals monitoring sufferers and helping families we **can** help in other ways. By promoting awareness of groups, services and other resources like blogs, books and websites which could help people – in addition to offering advice on alternative therapies they may find helpful in the meantime. These things may not be enough alone to help someone recover when they need specialist treatment and therapy, but they are better than nothing.

What is helpful?

This is equally if not more important than talking about the things which are harmful and damaging – because these are the things that can help people and increase the chances of eventually eliminating Anorexia for good.

Awareness of common causes

Being aware of things which can cause an eating disorder and therefore being on the lookout for any signs in someone vulnerable is key to identifying the early stages of Anorexia especially. Any sort of life trauma, for example a death in the family or a divorce can trigger an eating disorder. Having lots of difficult things going on and feeling helpless and out of control can be a trigger – as it was for me. Food is one of the easiest things in our lives available for us to control. Demanding parents, problems at school like bullying and low self-esteem are also all factors. Often it's the pressure we put on ourselves which causes us to be ill – rather than the pressure we're under from others – but this can certainly cause us to have unreasonably high expectations of ourselves.

The difficulty is that an eating disorder is such a secretive illness – so not only will the person be unlikely to tell you they have a problem (they may be at a stage where they don't think it *is* a problem), they might also be hiding what they are doing incredibly well.

Kindness and consideration – nurture the person inside

Compassion is key when it comes to Anorexia. I think that's because as I've said, whilst you present as somebody who isn't you, *you* are still there inside somewhere and you need kindness more than ever. Unfortunately Anorexia needs harsh, firm treatment and being separated from it doesn't at all feel good because you are so absorbed in all the cruel lies it's told you about yourself and about your life. Eventually though it is possible and you can squash that voice in your head.

To do this, I believe a balance is needed. A balance of targeted therapy and medical treatment which does involve a firm, uncompromising approach, and careful rediscovery of the personality, dreams and hope that may have been lost along the way through kindness and empathy.

Don't make assumptions - even if you don't understand. Instead make a point of being generous and friendly. Nurture the person inside. Remember that they are not the eating disorder, and the eating disorder is not them. They are separate entities.

The truth is, overcoming Anorexia is down to the individual. The responsibility lies with them; and only them. It's not something you can force them to do.

At the moment, they may not feel like life is worth living. That's not something you may feel able to change, but there are things that you can do to help them believe they are worthy of life just as anyone else is, and that they are capable and deserving of recovering and doing the things they want to.

Everyone has dreams. Likes, dislikes. We're all passionate about something. Anorexia makes you passionate about only one thing: controlling food. It supresses all the other things you loved doing and replaces your life with one which revolves around disappearing, physically and psychologically. But that person is still in there somewhere! They still want to do the things they always did. They have a warped perception of reality, and food will still be getting in the way of their dreams and day to day life. I still had dreams, (I wanted to live abroad by the beach) but in my dreams I lived off one cereal bar a day (which was perfectly acceptable to my completely irrational mind), and I weighed 6 stone (a massive 1.5 stone increase from where I was at the time, which seemed like a colossal amount to me at that point). For this reason it's also important to take unrealistic ideas of recovery and what 'being well and healthy' is into consideration. You might think a relatively small step is reasonable, but for them it might feel impossible, so be prepared to support a person with Anorexia more than you might have expected to initially. I often needed my parents to put things into perspective for me continually and reassure me constantly before I felt ready to take what was a very small step in the right direction, like eating a tiny piece of chocolate.

Treatment of Anorexia has to feel harsh and unpleasant to be effective. It's going to hurt the sufferer by default, because they wholly believe that what they are thinking is a product of their own mind and not a mental illness. This is why I think it's really important to show kindness too, whilst showing the eating disorder where to go.

The difficulty for parents and professionals alike is that you care about that person's health and want them to be well. But it's hard to decipher whether what they are telling you will make them happy or whether it actually makes the eating disorder happy and therefore contributes to their illness. Generally, what they say makes them happy is not what makes them healthy. Remember that it isn't them – it is the illness, and that there is still a person in there who has things (unrelated to food and exercise) which make them happy. Focus on these. Find, feed and encourage things that make that person happy which are completely unrelated to the eating disorder. For me, that was getting a pet, going for a walk, having a look round the garden centre, listening to music.

By likening Anorexia to a demon I am not waiving responsibility for how I behaved and what I put my family through. However I think it's important to separate the two when you look at someone you love who is suffering. Nurture the person inside them who is lost and screaming to get out – show the eating disorder no mercy, yet afford the person left love and attention. They are already at their lowest ebb, feeling helpless and worthless and guilty. By showing them they have something to live for, sharing with them their favourite activities, and encouraging them to look forward to the future, you help them grow and grow until eventually the demon inside their head grows smaller and smaller.

Whilst they are crying over the prospect of eating a meal, promise something fun at the end of it. Spend time with them, encourage them to spend time on themselves such as having a manicure or pedicure. One health worker Mary used to come and give me a manicure and hand massage every week; she brought in nail polish and olive oil and special hand cream just so that she could do that for me. One nurse, Joy, used to take me around the hospital grounds in my wheelchair and sit out in the sun with me. Joy always mentioned my smile and never the rest of me. She gave me hugs and genuinely wanted me to be well again. We would talk about the future, my plans, where I wanted to live and what I wanted to do. She always told me I had a beautiful smile; she was such a kind, positive, happy lady. These people made me feel like I was worth talking to and spending time with, whilst most other health professionals avoided me; they were used to dealing with physical illness and perhaps they had a fear of the unknown. They were angry and frustrated with me; they thought of me as a time waster. Of course my parents were incredible too. They would take me out for long wheelchair rides in the evenings too when they were allowed to visit, and when I was able to they would do lots of nice things with me so that slowly, I remembered who I was and why I was living. I remember those people and what they did for me so strongly and I am so thankful for their kindness and patience. The simple things they did made a huge difference and aided my complete recovery.

Small acts of kindness from strangers were also much appreciated. Where most people stared and avoided me, some smiled (a genuine smile, not a pity smile) and treated me like a human being – that meant a lot. In hospital, isolated and ignored by staff and the other patients (children who were more than likely frightened of me in my emaciated state), one day a group of student nurses were brought around the ward. They all huddled at the door as the tutor pointed at each of us in turn. Then they left. 5 minutes later, one of the nurses came back to the ward. She had come to tell me that she had just been through the same thing and was in the process of recovery; she showed me the downy hair on her arm that hadn't subsided yet. She told me everything was going to be okay. I can't tell you how grateful I was that she took the time to sneak away from her class to come and make me feel better, to let me know that I was not alone. She could have stayed with the class and moved on, but she saw me sitting there on my bed on my own so poorly and took the opportunity to bring me a little hope. It still means such a lot to me now that she did that, and if I knew who and where she was I would thank her for her small act of kindness that had a big impact on how I felt.

So you can see, you can make a huge difference to somebody simply by being kind. The smallest things had the biggest impacts – going shopping with my family my cousin making me do wheelies in the wheelchair – driving out in the evening sun listening to music. Being in nature, sitting with the sun on my face with Joy as we talked about the future.

I'm not a psychologist or doctor - I only speak from experience. But I believe that showing kindness whilst having to treat a mental illness so harshly and urgently is essential for the recovery of that person.

Helping sufferers to focus on something positive

In Tough Cookie I talk about one of the biggest catalysts for recovery being an ability to picture the future and look forward to positive things coming up, as well as identifying the good things you have in life right now. When you have Anorexia it can be hard for you to do this all by yourself, especially if you are consumed by a negative mind-set.

So it's really helpful for family, friends and carers to lend a helping hand. Make them feel important inside –encourage them to talk about what they like what they have to live for and give them practical examples of how they can start to do that right now. Look through magazines and at visual aids with them to make the goals tangible. More than anything this gives them hope indirectly without seeming like a conscious effort to combat Anorexia.

You can discover incredible things about yourself – what you want, what you're capable of and things you really enjoy when you have Anorexia. And just like any serious illness, having a near miss like that can make you really appreciate your life. Many incredible artists, writers and musicians emerged from the face of adversity.

Joy encouraged me to talk about the future and what I wanted in my life. She'd take me out in my wheelchair and we'd sit in a spot on the hospital grounds in the warm sun. She asked me what I was going to do when I was 'better'. Coming from her, I didn't see it as a loaded question or a trick – I saw that she was genuinely interested and cared about me. Together, we laid exciting plans for my life abroad by the sea. I'd talk excitedly about them to my parents who helped me to elaborate on my vision until it was looking grand and too good to forget or ignore. I still have those dreams today – and although I didn't know it at the time looking back I know that focusing on things like that helped me to want to recover.

I include visualisation tools and brainstorms in Tough Cookie which are easy to fill out and can help to fire up the person left inside by identifying what it is they truly desire – and realising that they can't have those things whilst they're still consumed by Anorexia.

Recognising signs of recovery and encourage and support

Recognising and noticing the early indications that somebody is starting to feel like they want to get better is really important. You can then encourage them and give them the support they will need to continue feeling strong enough to recover. People tend to talk about recovery as though it's a straight, steady upward road then you reach the top and you're 'better'. But actually it can be a long, hard slog filled with peaks and troughs and though there will be lots of good days and positive progress there might also be setbacks and relapses.

Signs of recovery are subtle but if you know what to look out for they're not too difficult to spot. When I had Anorexia, I had completely lost my personality. I spent most of the time silently staring into space – I only became animated if I was threatened with the prospect of having to face food. Any tiny spark of me in the midst of the way I presented whilst I was consumed with Anorexia was a sign of the 'me' inside showing through. Showing interest in things other than food or eating – talking about the future, laughing or even smiling at a joke. These are all things to look out for.

I also say in Tough Cookie that for me, guilt was one of the first signs that I was getting stronger. Anorexia is so powerful that it almost shuts out the guilt when you are throwing away a lovingly prepared meal or screaming at your loved ones. Feeling any element of guilt was certainly for me a turning point. Although at first it wasn't pleasant to feel so guilty but also feeling so helpless to do anything about hurting the people I loved, it was the start of my realisation and recovery.

With zero respect or love for myself (and no therapy to help me to change that), I recovered for the sake of my parents. We're always encouraged to do things for ourselves and only us, and that's undeniably true. In an ideal world, that's what would happen. But how many of us really do things for ourselves? We tend to have more love in our hearts for others than we do for ourselves. I think it's important recognise that people's motive for recovery might not be 'healthy' or 'ideal'; but if it's a motive it's a motive – let it be and once they are on their way they will be more responsive to therapy, more open to perhaps feeling better about themselves and that *they* are worth a healthy, happy life. The motive isn't so important at first – what is crucial is that that person feels ready to make a change and live their lives and whatever the circumstances that should be encouraged.

Supporting family and friends

Watching your loved one turn into someone you don't recognise – physically and personality-wise is an upsetting experience that simply cannot be trivialised. They've lost the person they thought they knew – and desperately want them back.

I credit my parents with my successful recovery from Anorexia – but my mum has always said she couldn't have done it without the support of a therapist they had at the hospital who was actually very generous and caring. She gave them the tools and advice and listened to them when they became dejected and downhearted if I wasn't doing too well. Recovery is a struggle – and more and more families are dealing with Anorexia alone because of a lack of places on specialist wards.

As I've said before, I know we can't change this – but what we *can* change is the fact that lots of parents and carers go unaided in their fight against Anorexia.

The pressure on that individual at the moment as a parent, carer or friend is enormous. They feel responsible for this person and their recovery. It's the hardest thing in the world to helplessly watch someone you love slowly killing themselves, feeling powerless to do anything about it.

As I say throughout this book, I believe my parents were instrumental in pulling me through Anorexia and ensuring my long term recovery – in fact I genuinely believe that without them I wouldn't be here today. But they were only able to do that because they did have some outside support – not a lot, but some.

One thing I've noticed when speaking to parents is that the ones who try to take care of themselves in order support their families have stronger, more resilient children as a result. For me it's so important for loved ones to understand what to do and what not to do in such a distressing and confusing time, especially when there's often little support for families who are going through it with their loved ones. Whilst the experience was traumatic for me, I was so wrapped up in my own suffering that I think it was actually harder on my parents, who were fully present and so had to witness my self-destruction and had to struggle and fight for my survival and recovery – and with me having no help or support from anyone else, that made it a million times harder for them. The dynamics in our house were very strange, because on the one hand my parents naturally (like most parents!) wanted to feed me and see me happy and healthy, and on the other I understood that and to know you are hurting your loved ones but feeling powerless to do anything about it is horrendous. Only now I fully appreciate the fact that they worked as a team and didn't make me feel guilty or punish me.

Luckily, they had access to the therapist who had gone out of her way to research Anorexia and armed with a few tools helped them to get through it and get me through it, too. Not all parents have access to that kind of support – but as I say frequently in this book, even small things make a difference. Perhaps you can refer them to a counselling service, or recommend a book they could buy – maybe you know there are support groups or forums where they can speak to other parents in the same position. If nothing else, show them kindness and support. Let them know that they are going through something tough, and that they are doing really well to keep everything together.

Helping those who ask for it – in any way you can

I've already covered this partially in 'Lack of Urgency' – but you'll probably understand by now that for me, 'help' comes in all sorts of forms. Just taking the time to look whether there are any resources out there or handing someone the number for a free counselling charity – whether they use it or not – shows you care – and that can be more helpful than you might know.

If someone has asked for help, then the biggest thing you can do knowing how serious Anorexia can be is do your very best to get them that help. Treat that person as a human being and talk to them directly at all times – research local centres and referral systems and if there is a waiting list, don't leave them high and dry – make sure that you're checking in on them regularly and see if there are any interim services they can access whilst they wait.

9. TREATMENT NOTES AND GUIDELINES

How to talk to a young person with Anorexia

It can be difficult to know how to approach someone with Anorexia, as they are often volatile and don't wish to discuss what is going on. When you have Anorexia you strongly dislike being 'accused' of having anything wrong with you, because you can't differentiate between your eating disorder and your own mind. This can make broaching the subject of treatment and care tricky.

How to discuss Anorexia with a young person and their parents

Talking to an individual with Anorexia can, as discussed above, prove difficult. This makes it especially important to recognise the role of parents and family members in recovery, and to be able to speak to them about the condition of their loved one in a way which they will understand and take on board.

As with lots of distressing mental or physical illnesses, parents are likely to be anxious, frustrated and upset. They may also feel helpless and could be annoyed about the situation as a whole and how it is being dealt with. As Anorexia has both serious physical and mental concerns, reassuring them calmly and explaining the next steps can go a long way to helping them to feel as though things may get better. Ensuring that parents feel supported themselves and are aware of what is going on regarding their treatment is important. If treatment isn't available, leave them with appropriate information and details of groups and organisations who may be able to help in the meantime. Offer general advice on how they can make things better at home for them and their child, and discourage blaming, shaming, denial and pandering to the eating disorder, explaining that all of these things are likely to worsen their child's condition.

Dispelling common myths (such as the ones discussed earlier in this book) may also be helpful – especially if parents have little or no understanding and are particularly angry or unhelpful when it comes to dealing with their child.

What to do if you suspect someone has Anorexia

Of course there are set guidelines for professionals at all levels to follow when dealing with eating disorders – so I won't discuss that here in this book. However here are a few general pointers which will be elaborated on a little in the next chapter, FOCUS:

- *Treat the case with urgency*
- *Refer if appropriate (or possible)*
- *Don't underestimate – employ strict observation*
- *Check the individual's support network*
- *Look for other resources and organisations*

10. FOCUS

I created this acronym to hopefully help you to remember the key messages contained within this book when you come into contact with a person with Anorexia, especially somebody younger. FOCUS is designed to bring together all the smaller points in this book into a few main essentials which are easier to remember, to be considered when treating an individual with Anorexia both initially and during treatment.

F - Focus

O - Observation

C - Consideration

U - Urgency

S - Support

Focus: Help the person to focus on something positive. Whether you are able to do this will depend on the role you are in – as a GP or doctor you may not have sufficient contact with the patient to do so, but as a HCA or nurse you will probably have the opportunity to do this, however briefly. As I've covered earlier in the book, some of the small, often seemingly insignificant things professionals did and said during my stay in hospital made a big difference to me, good and bad.

Observation: I can't stress how important observation is – both medically-speaking but also simply in terms of watching the individual as closely as possible. Anorexia is devious and it becomes very easy to know how to evade being seen hiding food, exercising and purging. This can't be underestimated, so constant observation is essential is possible. Pay attention to their behaviour and watch out for suspicious or unusual activities or responses, such as going to the toilet straight after a meal, or defensiveness over a bag or bin nearby them which they may have been using to dispose of or hide food.

Consideration: Although you may be busy, stressed or pushed for time, try to show consideration when dealing with the individual, especially if they are being frustrating or appear to be behaving in a difficult manner. They are likely to be frightened, anxious and even angry, and may have other acute mental health conditions as well as Anorexia, so an understanding of how they are feeling is important when treating them so that ultimately they are more responsive to what you say and do. A little kindness and a caring attitude goes a very long way when you are feeling isolated, alone and are mentally and physically unwell.

Urgency: Anorexia is serious mentally but it can also be deadly physically, and an individual's condition can deteriorate quickly. It's also difficult to know how long they have had Anorexia for and exactly how they have been behaving, how little they have been eating/drinking, any other factors in play – as they're likely not to be truthful when asked. In a GP situation urgency when referring and checking the current state of the patient is key if possible. In a hospital situation all medical tests should be done as quickly as possible and the person should not be left alone – as above. A referral should be made if appropriate depending on their current condition and if the individual is at a very low weight they should be monitored and looked after accordingly.

Support: Support the patient by trying to keep them in the loop regarding their treatment and care and offering advice and support in a calm and caring way. Also support family and friends or anyone around them by explaining what the process may be and offering information regarding other resources, services, charities or organisations they may be able to access to help them to help their family member.

11. THE LAST WORD

I hope that this book has provided an interesting insight into the mind of somebody who has been through Anorexia. Whilst what happened to me was shocking and is now probably very rare, we still struggle to provide appropriate care for those with eating disorders in this country for a number of reasons and lots of people still have bad experiences, which is unfortunate.

I'm convinced that collectively we all have a role to play in reducing the rate of eating disorders. However complex and dangerous this illness is, **you can make a difference**. A kind word or a thoughtful gesture can mean a lot. Feeling like you have someone who understands is so very important – and can even instil hope in someone or inspire them to get better – especially when it comes from someone in a professional position. Small things make a big difference – and whilst I know that we can't change circumstances directly and I can't bring change single-handedly, I believe that arming professionals with the tools with which to recognise and treat eating disorders is a valuable first step.

As with any mental illness, it's important to treat sufferers with humility and understanding. Any of us could develop an eating disorder at any time in our lives, so my final message is to please always try to treat these vulnerable people as you would wish to be treated yourself.

12. CONTACT, FURTHER READING AND RESOURCES

Further reading and resources

This list is not exhaustive but contains websites, books and organisations you can recommend to someone with Anorexia or their family.

Charities and Organisations

Mind – Mind have local centres all over the UK which individuals can visit and use to access support both within and outside the NHS. Relatives can also source counselling and find help sheets and resources which they may find useful.

ANAD - This is an American site but the resources and articles on it are helpful for anyone suffering with Anorexia and their families.

Beacon – Beacon provide counselling services which are often free (but are sometimes subsidised or available for a small fee). This can be helpful in the interim when a referral is not possible, and for family and friends who need someone to talk to.

SANE – SANE offers resources for people with a range of mental health issues including eating disorders.

Support Line - www.supportline.co.uk

Tough Cookie - www.toughcookieblog.co.uk

NHS site – www.nhs.uk

It's likely you'd recommend this as a reference anyway, but I included it here as it is worth noting there is plenty of information on eating disorders and Anorexia in particular on the site for sufferers and family members.

Books

As well as this book I have written several others. Tough Cookie was written specifically for people with eating disorders (Anorexia in particular, but also Bulimia) to share my experience and the things which helped me to recover in a positive way. Eating Disorders: Parent Handbook has been designed to help parents to help their child, as mine were instrumental in my recovery. You can find details of these books on my website as above.

Contact me

I'm passionate about helping professionals and individuals alike to understand eating disorders and I promote positivity and empathy to inspire recovery. If you would like more information on my books, or would like to talk to me regarding a keynote, course or training package, then please get in touch.

info@toughcookieblog.co.uk

www.ingramcontent.com/pod-product-compliance
Lightning Source LLC
Chambersburg PA
CBHW072232170526
45158CB00002BA/864